Editor
Leasha Taggart

Editorial Manager
Karen J. Goldfluss, M.S. Ed.

Editor in Chief
Sharon Coan, M.S. Ed.

Illustrator
Chandler Sinnott

Cover Artist
Jessica Orlando

Creative Director
Elayne Roberts

Creative Coordinator
Denice Adorno

Product Manager
Phil Garcia

Imaging
James Edward Grace
Ralph Olmedo, Jr.

Publishers
Rachelle Cracchiolo, M.S. Ed.
Mary Dupuy Smith, M.S. Ed.

Ho
Add & Subtract

Grade 2

Author

Mary Rosenberg

Teacher Created Materials, Inc.
6421 Industry Way
Westminster, CA 92683
www.teachercreated.com

ISBN-1-57690-943-3

Reprinted, 2000
Made in U.S.A.

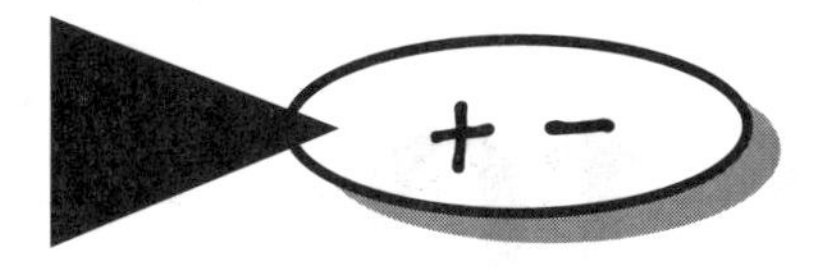

Table of Contents

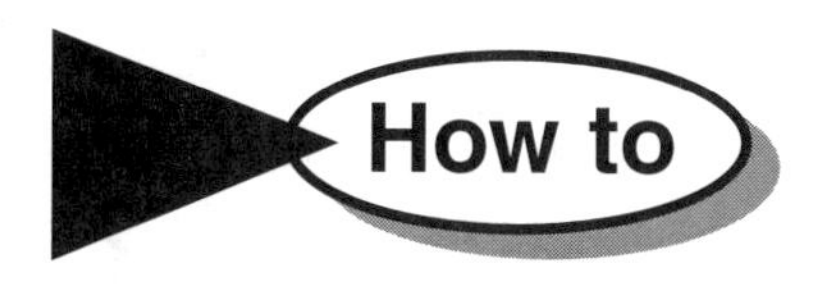

Use This Book

A Note to Teachers and Parents

Welcome to the "How to" math series! You have chosen one of over two dozen books designed to give your children the information and practice they need to acquire important concepts in specific areas of math. The goal of the "How to" math books is to give children an extra boost as they work toward mastery of the math skills established by the National Council of Teachers of Mathematics (NCTM) and outlined in grade-level scope and sequence guidelines. The NCTM standards encourage children to learn basic math concepts and skills and apply them to new situations and to real-world events. The children learn to justify their solutions through the use of pictures, numbers, words, graphs, and diagrams.

The design of this book is intended to allow it to be used by teachers or parents for a variety of purposes and needs. Each of the units contains one or more "How to" pages and two or more practice pages. The "How to" section of each unit precedes the practice pages and provides needed information such as a concept or math rule review, important terms and formulas to remember, or step-by-step guidelines necessary for using the practice pages. While most "How to" pages are written for direct use by the children, in some lower-grade level books these pages are presented as instructional pages or direct lessons to be used by a teacher or parent prior to introducing the practice pages. In this book, the "How to" pages detail the concepts that will be covered in the pages that follow as well as how to teach the concept(s). Many of the "How to" pages also include "Learning Notes" and "Teaching the Lesson." The practice pages review and introduce new skills and provide opportunities for the children to apply the newly acquired skills. Each unit is sequential and builds upon the ideas covered in the previous unit(s).

About This Book

How to Add & Subtract: Grade 2 presents a comprehensive overview of addition and subtraction of whole numbers on a level appropriate for students in grade 2. The clear, simple, readable instruction pages for each unit make it easy to introduce and teach basic addition and subtraction to children with little or no background in the concepts.

The use of manipulatives as visualization tools is encouraged. In addition, if children have difficulty with a specific concept or unit within this book, review the material and allow them to redo troublesome pages. Since concept development is sequential, it is not advisable to skip much of the material in the book. It is preferable that children find the work easy and gradually advance to the more difficult concepts at a comfortable pace.

The following skills are introduced in this book:

- adding and subtracting to 18
- adding and subtracting 2- and 3-digit numbers with and without regrouping
- place value to 3 places (hundreds, tens, and ones)
- problem-solving skills and strategies
- estimation
- number sense
- communication of math ideas through pictures, numbers, words, charts, and labels
- developing math reasoning skills
- adding 3 numbers
- patterns
- measurement
- using charts for information
- collecting data and making charts based on the data
- fractions
- using calculators
- using the computer to reinforce subtraction

Standards

The units in this book are designed to match the suggestions of the National Council of Teachers of Mathematics (NCTM). They strongly support the learning of addition and subtraction and other processes in the context of problem solving and real-world applications. Use every opportunity to have students apply these new skills in classroom situations and at home. This will reinforce the value of the skill as well as the process. The activities in this book match the following NCTM standards:

Problem Solving

The children develop and apply strategies to solve problems, verify and interpret results, sort and classify objects, and solve word problems.

Communication

The children are able to communicate mathematical solutions through manipulatives, pictures, diagrams, numbers, and words. Children are able to relate everyday language to the language and symbols of math. Children have opportunities to read, write, discuss, and listen to math ideas.

Reasoning

Children make logical conclusions through interpreting graphs, patterns, and facts. The children are able to explain and justify their math solutions.

Connections

Children are able to apply math concepts and skills to other curricular areas and to the real world.

Number Sense and Numeration

Children learn to count, label, and sort collections as well as learn the basic math operations of addition and subtraction.

Concepts of Whole Number Operations

Children develop an understanding for the operation (addition, subtraction) by modeling and discussing situations relating math language and the symbols of operation (+, –) to the problems being discussed.

Geometry and Spatial Sense

Children are able to describe, model, draw, and classify shapes, and relate geometric ideas to number and measurement ideas.

Other Standards

Children use **measurement** (time, distance, weight, volume, area, etc.) and estimates of measurement in problem solving and in everyday situations. Children learn about **statistics** and **probability** as they collect and organize data into graphs, charts, and tables.

Children develop concepts of **fractions** through the use of pattern blocks. Children are able to model, explain, and develop competency in basic facts, **mental computation**, and **estimation** techniques.

Children will be able to recognize, describe, and extend a variety of **patterns** as well as represent and describe math relationships.

1 How to ••••••••••••• Add and Subtract Three Single-Digit Numbers

Learning Notes

In this unit children practice adding and subtracting three single-digit numbers. They are introduced to the concept of "doubles" (two of the same number being added together) and to the addition strategy of "doubles + 1."

Children also apply their knowledge of adding and subtracting three numbers to word problems.

Materials

- counters (multilink cubes, bottle caps, craft sticks, buttons, coins, etc.)
- 3 cups (or circles drawn on paper) in which to place manipulatives when adding three numbers

Teaching the Lesson

Before the lesson, have the children practice adding three different numbers through the use of manipulatives.

For example, present the following information:

Three kittens were playing. Two kittens joined them. Four more kittens came to play. How many kittens are there in all?

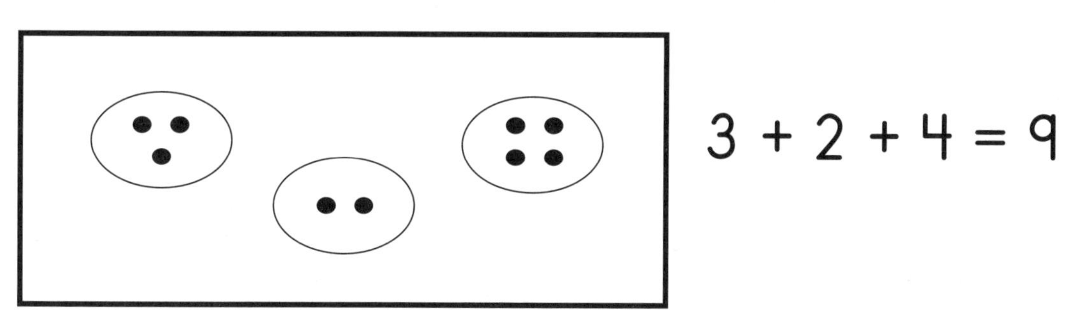

$3 + 2 + 4 = 9$

There were 10 stamps. Henry used 3 to mail letters. He gave 2 to his mom. How many stamps does Henry have left?

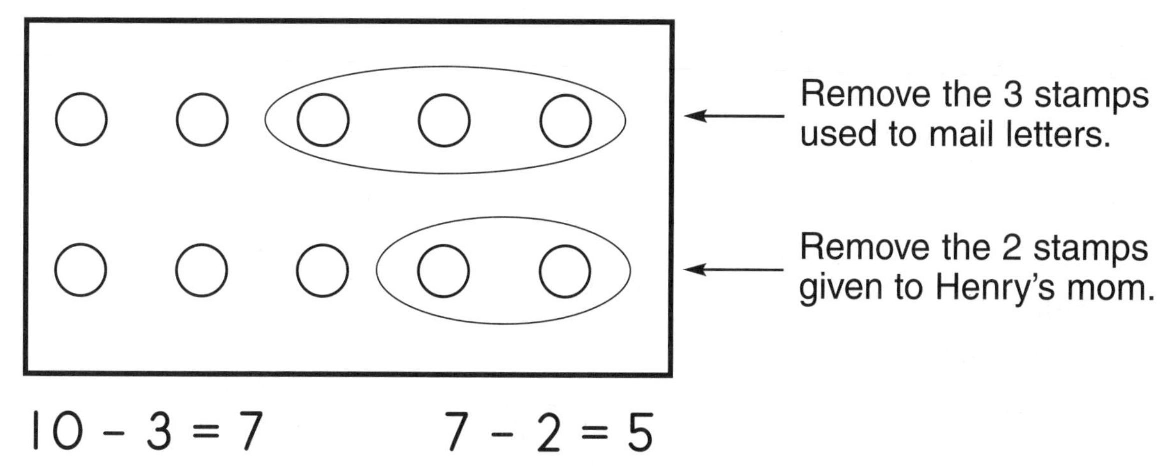

$10 - 3 = 7$ $7 - 2 = 5$

Model for the children how to use the "box" on the Adding Three Numbers practice page (page 7) to "hold" the total value of the first two numbers while they "count on" to add the third number.

Adding Doubles + 1

Look at each pair of number facts. The first addition problem uses doubles (2 of the same number). The second addition problem uses numbers that are "neighbors" (a double + 1 more). Solve each addition problem.

1. $1 + 1$ **2.** $1 + 2$	**3.** $8 + 8$ **4.** $8 + 9$	**5.** $7 + 7$ **6.** $7 + 8$
7. $5 + 5$ **8.** $5 + 6$	**9.** $3 + 3$ **10.** $3 + 4$	**11.** $9 + 9$ **12.** $9 + 10$
13. $4 + 4$ **14.** $4 + 5$	**15.** $0 + 0$ **16.** $0 + 1$	**17.** $6 + 6$ **18.** $6 + 7$

Mental Math

Solve each problem.

19. 4 + 4 + 5 = ______

20. 1 + 1 + 2 = ______

21. 5 + 5 + 6 = ______

22. 3 + 3 + 4 = ______

23. 0 + 0 + 1 = ______

24. 2 + 2 + 3 = ______

1 Practice • • • • • • • • • • • • • • Adding Three Numbers

When adding three numbers, add the first two numbers together. Write that answer in the box and add the third number by "counting on." Write the answer below the line. The first one has been done for you.

1.
$$\begin{array}{r} 3 \\ 6 \\ +\ 8 \\ \hline 17 \end{array} > \boxed{9}$$

2.
$$\begin{array}{r} 9 \\ 3 \\ +\ 6 \\ \hline \end{array} > \square$$

3.
$$\begin{array}{r} 5 \\ 5 \\ +\ 5 \\ \hline \end{array} > \square$$

4.
$$\begin{array}{r} 4 \\ 0 \\ +\ 2 \\ \hline \end{array} > \square$$

5.
$$\begin{array}{r} 3 \\ 5 \\ +\ 7 \\ \hline \end{array} > \square$$

6.
$$\begin{array}{r} 4 \\ 2 \\ +\ 9 \\ \hline \end{array} > \square$$

7.
$$\begin{array}{r} 7 \\ 0 \\ +\ 1 \\ \hline \end{array} > \square$$

8.
$$\begin{array}{r} 2 \\ 6 \\ +\ 1 \\ \hline \end{array} > \square$$

9.
$$\begin{array}{r} 5 \\ 6 \\ +\ 5 \\ \hline \end{array} > \square$$

10.
$$\begin{array}{r} 5 \\ 3 \\ +\ 4 \\ \hline \end{array} > \square$$

11.
$$\begin{array}{r} 8 \\ 6 \\ +\ 3 \\ \hline \end{array} > \square$$

12.
$$\begin{array}{r} 7 \\ 2 \\ +\ 4 \\ \hline \end{array} > \square$$

Write the missing numbers.

13.
$$\begin{array}{r} 5 \\ 6 \\ +\ \square \\ \hline 14 \end{array}$$

14.
$$\begin{array}{r} 4 \\ 4 \\ +\ \square \\ \hline 12 \end{array}$$

15.
$$\begin{array}{r} 7 \\ 3 \\ +\ \square \\ \hline 10 \end{array}$$

16.
$$\begin{array}{r} 0 \\ 9 \\ +\ \square \\ \hline 15 \end{array}$$

Addition and Subtraction Word Problems

Read each word problem. Write the math problem and the answer.

1. Cheryl has 3 goldfish, 4 yellow fish, and 6 green fish. How many fish does Cheryl have in all?

Cheryl has 13 fish in all.

$$\begin{array}{r} 3 \\ 4 \\ +\ 6 \\ \hline 13 \end{array}$$

2. Marcus made a headband. He used 7 green feathers, 8 blue feathers, and 1 red feather. How many feathers did Marcus use in all?

Marcus used ________ feathers in all.

$$\begin{array}{r} \square \\ \square \\ +\ \square \\ \hline \square \end{array}$$

3. Jasmine used beads to make a necklace. She used 6 brown beads, 5 pink beads, and 9 black beads. How many beads did Jasmine use in all?

Jasmine used ________ beads in all.

$$\begin{array}{r} \square \\ \square \\ +\ \square \\ \hline \square \end{array}$$

4. Angelo likes to play marbles. He has 3 that are cat eyes, 5 that are swirled, and 6 that are solid blue. How many marbles does Angelo have in all?

Angelo has ________ marbles in all.

$$\begin{array}{r} \square \\ \square \\ +\ \square \\ \hline \square \end{array}$$

5. Paula had 18 seeds. She gave 7 to her sister and 6 to her brother. How many seeds does Paula have left?

Paula has 5 seeds left.

$$\begin{array}{r} 18 \\ -\ 7 \\ \hline 11 \\ -\ 6 \\ \hline 5 \end{array}$$

6. Nick had 15 yo-yos. He lost 3 on his way to school and gave 8 away. How many yo-yos does Nick have left?

Nick has ________ yo-yos left.

$$\begin{array}{r} \square \\ -\ \square \\ \hline \square \\ -\ \square \\ \hline \square \end{array}$$

7. Spot had 14 balls. He buried 2 in the backyard and 6 in the front yard. How many balls did Spot not bury?

Spot did not bury ________ balls.

$$\begin{array}{r} \square \\ -\ \square \\ \hline \square \\ -\ \square \\ \hline \square \end{array}$$

8. Wilma had 17 letters in her alphabet soup. She ate all of the vowels (a, e, i, o, u) and 8 consonants. How many letters does Wilma have left?

Wilma has ________ letters left.

$$\begin{array}{r} \square \\ -\ \square \\ \hline \square \\ -\ \square \\ \hline \square \end{array}$$

2 How to ••••• Make Number and Shape Patterns

Learning Notes

The children learn that patterns can be found just about anywhere in the world. They use different items, such as letters, numbers, colors, and shapes to create patterns. Children also learn that letters can be used to describe a pattern.

Materials

- pattern blocks (or several different objects that can be used to create a pattern)
- chalkboard or a piece of paper

Teaching the Lesson

Before teaching the lesson, ask the children if they know what a pattern is.

Clap and snap your hands to create a pattern. For example, use the pattern "clap, clap, snap, clap, clap, snap." Have the children continue the pattern with you. Ask the children how they would describe the pattern to someone else. At this point, introduce the idea that patterns can be labeled with letters. The "clap, clap, snap" pattern would be labeled "AAB." Use the chalkboard or a piece of paper to model how to label the pattern.

Have the children make up patterns of their own to share with their classmates.

Go over the following practice pages with the children.

Naming the Pattern (page 10): Use numbers to create familiar patterns but with unfamiliar numbers. For example, counting by 2's but start with the number 31.

Identifying the Pattern (page 11): Use shapes to make patterns and pictures.

Making a Pattern (page 12): The children use pattern blocks to create a variety of patterns. See the examples below.

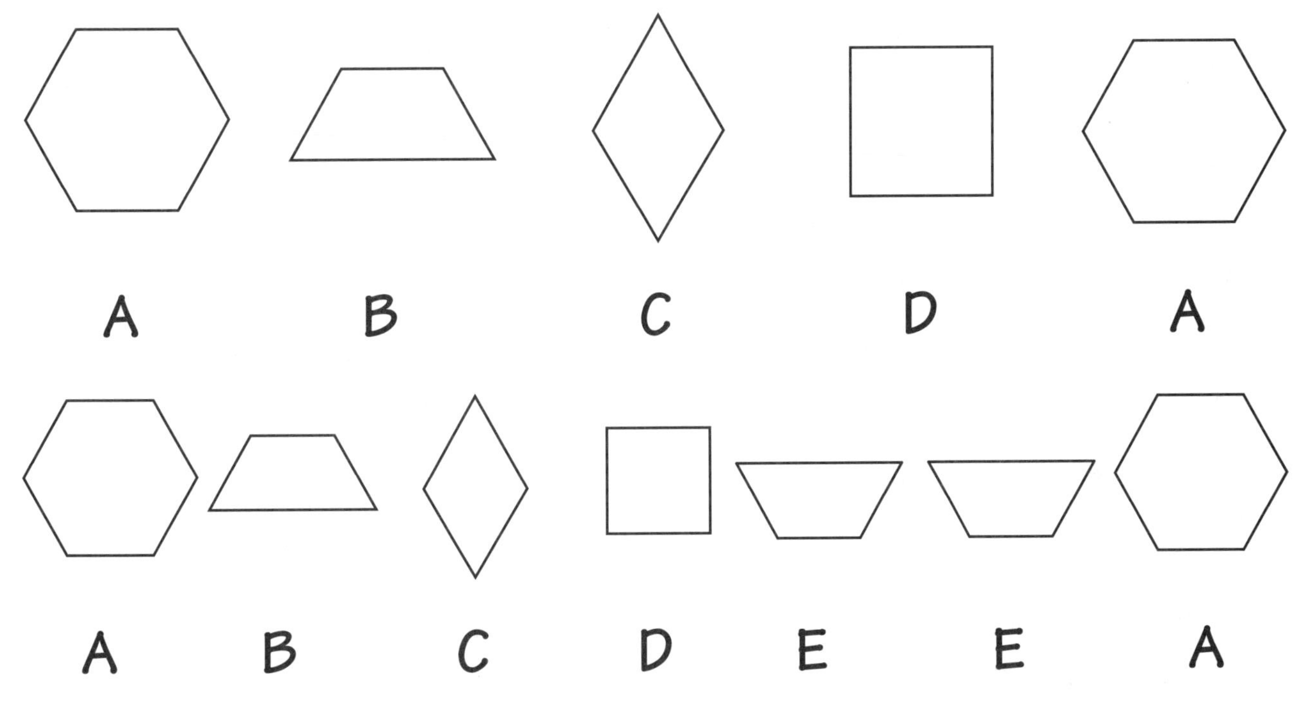

2 Practice •••••••••••••••••• Naming the Pattern

Complete the patterns.

1. 8, 10, 12, _____, _____, _____, _____, _____
2. 31, 33, 35, _____, _____, _____, _____, _____
3. 25, 27, 29, _____, _____, _____, _____, _____
4. What is the counting pattern?____________________

---

5. 15, 20, 25, _____, _____, _____, _____, _____
6. 71, 76, 81, _____, _____, _____, _____, _____
7. 47, 52, 57, _____, _____, _____, _____, _____
8. What is the counting pattern?____________________

---

9. 50, 60, 70, _____, _____, _____, _____, _____
10. 5, 15, 25, _____, _____, _____, _____, _____
11. 43, 53, 63, _____, _____, _____, _____, _____
12. What is the counting pattern?____________________

---

13. 50, 100, 150, _____, _____, _____, _____, _____
14. 130, 180, 230, _____, _____, _____, _____, _____
15. 610, 660, 710, _____, _____, _____, _____, _____
16. What is the counting pattern?____________________

---

17. 300, 400, 500, _____, _____, _____, _____, _____
18. 110, 210, 310, _____, _____, _____, _____, _____
19. 275, 375, 475, _____, _____, _____, _____, _____
20. What is the counting pattern?____________________

Complete each pattern. Use letters to label each pattern. The first pattern has been completed for you.

1. 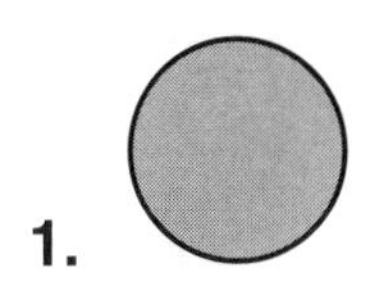 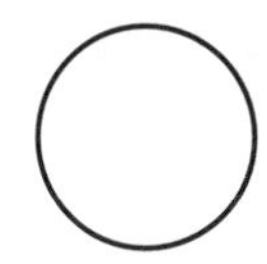  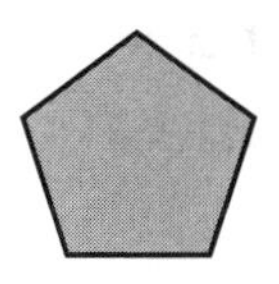  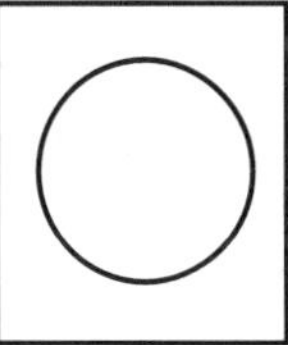 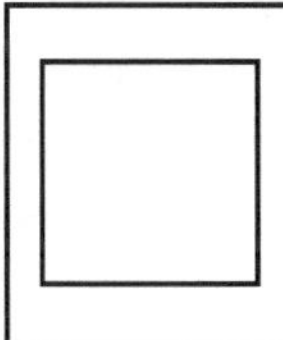 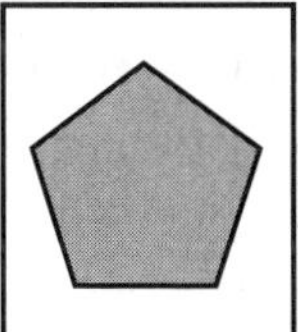

A B C D A B C D

---

2.       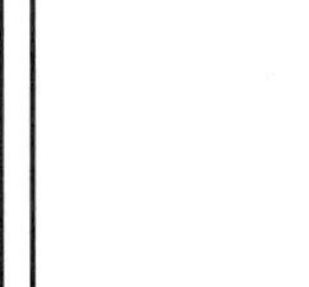

___ ___ ___ ___ ___ ___ ___ ___

---

3.       

___ ___ ___ ___ ___ ___ ___ ___

---

4.        

1, 1 2, 1 2 3, 1

___ ___ ___ ___ ___ ___ ___ ___ ___ ___

---

5. On the back of this page, make a pattern of your own.

Pick 4 different pattern blocks to make a pattern. Draw the pattern you made. Use letters to describe the pattern. You can use more than 4 pattern blocks, but only 4 different shapes to make the pattern. For example, square, hexagon, diamond, and trapezoid are 4 different shapes.

Using the same pattern blocks, arrange the blocks in a new pattern. Draw the pattern you made. Use letters to describe the pattern.

Using the same pattern blocks, arrange the blocks in yet another new pattern. Draw the pattern you made. Use letters to describe the pattern.

Write 2 sentences telling about the patterns you made and anything that you noticed.

3 How to • • • • • • • • • • • Add and Subtract Two-Digit Numbers Without Regrouping

Learning Notes

Children learn to add and subtract two-digit numbers without regrouping and to apply the concept of "tens" and "ones" to math problems.

Materials

- sets of tens manipulatives: craft sticks, straws, or coffee stirrers in sets of 10 (rubber banded together); multilink cubes or connecting chains (snapped together to make tens); beans or small candies placed in sets of 10 in paper cups. Left loose as individual pieces, any of these items represent ones.
- 3" x 5" (8 cm x 13 cm) index cards numbered 1–100
- Tens and Ones Mat, which can be drawn on a piece of paper, as shown in the example on the right.

tens	ones

Teaching the Lesson

Before beginning the activities, show the children a number written on an index card and have the children make the number using the math manipulatives and the Tens and Ones Mat. Have the children make several numbers in this manner.

Model how to use the Tens and Ones Mat and manipulatives to make different addition and subtraction problems without regrouping. (See the examples below.)

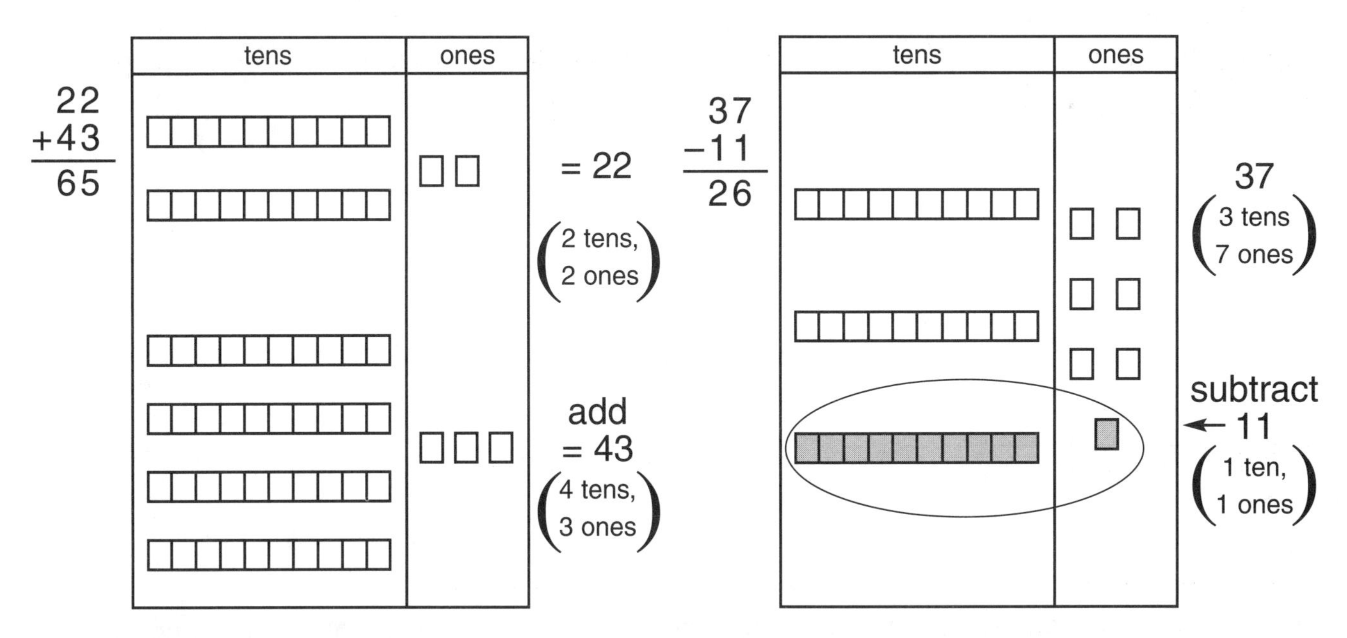

Have the children complete the unit activities. Remind the children to always start on the ones side whenever they add or subtract.

Learning Tip

For children who are having difficulty with this procedure, give them a self-sticking note. Place the note over the tens side of the math problem and do the adding/subtracting on the ones side and record the answer. Remove the note and do the adding/subtracting on the tens side and record the answer.

3 Practice • • • • • • • • • • • Adding Two-Digit Numbers Without Regrouping

When adding two-digit numbers, remember to always start on the ones side first.

1. tens | ones
$$\begin{array}{r} 20 \\ +46 \\ \hline \end{array}$$

2. tens | ones
$$\begin{array}{r} 70 \\ +29 \\ \hline \end{array}$$

3. tens | ones
$$\begin{array}{r} 30 \\ +30 \\ \hline \end{array}$$

4. tens | ones
$$\begin{array}{r} 37 \\ +50 \\ \hline \end{array}$$

5. tens | ones
$$\begin{array}{r} 29 \\ +30 \\ \hline \end{array}$$

6. tens ones
$$\begin{array}{r} 26 \\ +32 \\ \hline \end{array}$$

7. tens ones
$$\begin{array}{r} 73 \\ +25 \\ \hline \end{array}$$

8. tens ones
$$\begin{array}{r} 51 \\ +41 \\ \hline \end{array}$$

9. tens ones
$$\begin{array}{r} 18 \\ +31 \\ \hline \end{array}$$

10. tens ones
$$\begin{array}{r} 27 \\ +10 \\ \hline \end{array}$$

11. tens ones
$$\begin{array}{r} 45 \\ +41 \\ \hline \end{array}$$

12. tens ones
$$\begin{array}{r} 76 \\ +12 \\ \hline \end{array}$$

13. tens ones
$$\begin{array}{r} 16 \\ +22 \\ \hline \end{array}$$

14. tens ones
$$\begin{array}{r} 40 \\ +49 \\ \hline \end{array}$$

15. tens ones
$$\begin{array}{r} 71 \\ +20 \\ \hline \end{array}$$

16. Write the answers to each problem in order from smallest to greatest.

_____, _____, _____, _____, _____, _____, _____, _____, _____, _____, _____, _____, _____, _____, _____

Use the numbers from problem 16 to complete the activities below.

17. Draw a circle around the numbers that have the largest number in the tens place.

18. Draw a rectangle around the numbers that have the largest number in the ones place.

19. Draw a triangle around the numbers that have the smallest number in the tens place.

20. Draw a square around the numbers that have the smallest number in the ones place.

3 Practice • • • • • • • • Subtracting Two-Digit Numbers Without Regrouping

When subtracting two-digit numbers, remember to always start on the ones side first.

1.

tens	ones
2	0
− 1	0

2.

tens	ones
3	6
− 3	5

3.

tens	ones
3	9
− 2	7

4.

tens	ones
8	5
− 3	5

5.

tens	ones
7	8
− 4	8

6. tens ones
$$\begin{array}{r} 83 \\ -52 \\ \hline \end{array}$$

7. tens ones
$$\begin{array}{r} 80 \\ -10 \\ \hline \end{array}$$

8. tens ones
$$\begin{array}{r} 97 \\ -64 \\ \hline \end{array}$$

9. tens ones
$$\begin{array}{r} 78 \\ -15 \\ \hline \end{array}$$

10. tens ones
$$\begin{array}{r} 25 \\ -14 \\ \hline \end{array}$$

11. tens ones
$$\begin{array}{r} 89 \\ -41 \\ \hline \end{array}$$

12. tens ones
$$\begin{array}{r} 91 \\ -71 \\ \hline \end{array}$$

13. tens ones
$$\begin{array}{r} 88 \\ -28 \\ \hline \end{array}$$

14. tens ones
$$\begin{array}{r} 94 \\ -73 \\ \hline \end{array}$$

15. tens ones
$$\begin{array}{r} 81 \\ -61 \\ \hline \end{array}$$

16. Circle all of the answers that are odd numbers. (Odd numbers are numbers that have a 1, 3, 5, 7, or 9 in the ones place.)

17. Draw a square around all of the answers that are even numbers. (Even numbers are numbers that have a 0, 2, 4, 6, or 8 in the ones place.)

18. When subtracting an odd number from another odd number, is the answer an odd number or an even number? ________________

19. When subtracting an odd number from an even number, is the answer an odd number or an even number? ________________

3 Practice • • • • • • • • • • • • • Adding and Subtracting Two-Digit Numbers

Solve each math problem. Write the letter that goes with each answer on the line at the bottom of the page.

A	H	I	M	R	S	T
22	39	15	46	72	30	60

1. $\begin{array}{r} 66 \\ -\ 51 \\ \hline \end{array}$

2. $\begin{array}{r} 65 \\ -\ 43 \\ \hline \end{array}$

3. $\begin{array}{r} 88 \\ -\ 42 \\ \hline \end{array}$

4. $\begin{array}{r} 11 \\ +\ 11 \\ \hline \end{array}$

5. $\begin{array}{r} 15 \\ +\ 31 \\ \hline \end{array}$

6. $\begin{array}{r} 97 \\ -\ 75 \\ \hline \end{array}$

7. $\begin{array}{r} 70 \\ -\ 10 \\ \hline \end{array}$

8. $\begin{array}{r} 27 \\ +\ 12 \\ \hline \end{array}$

9. $\begin{array}{r} 65 \\ -\ 35 \\ \hline \end{array}$

10. $\begin{array}{r} 10 \\ +\ 50 \\ \hline \end{array}$

11. $\begin{array}{r} 48 \\ -\ 26 \\ \hline \end{array}$

12. $\begin{array}{r} 61 \\ +\ 11 \\ \hline \end{array}$

The secret message is:

___   ___ ___   ___   ___ ___ ___ ___   ___ ___ ___ ___ !

1   2 3   4   5 6 7 8   9 10 11 12

4 How to • • • • • • • • • • • • • • Regroup when Adding

Learning Notes

The children develop the concept of "tens" and "ones," sort objects into groups of ten, and learn to regroup.

Materials

- sets of tens manipulatives: craft sticks, straws, or coffee stirrers in sets of 10 (rubber banded together); multilink cubes or connecting chains (snapped together to make tens); beans or small candies placed in sets of 10 in paper cups. Left loose as individual pieces, any of these items represent ones.
- Tens and Ones Mat which can be drawn on a piece of paper, as shown in the example on the right

tens	ones

Teaching the Lesson

Take a handful of beans (or the manipulative item being used) and model for the children how 10 beans can make a set by placing the beans in a small cup. Each set of 10 beans is placed on the tens side. If there aren't enough beans left to make a 10, those beans stay on the ones side.

Have the children practice doing this procedure several times.

Using Place Value (page 18): Place a handful of beans and lay the beans on the table. Take a second handful of beans and lay the beans next to the first handful but in a separate pile.

First handful

Second handful

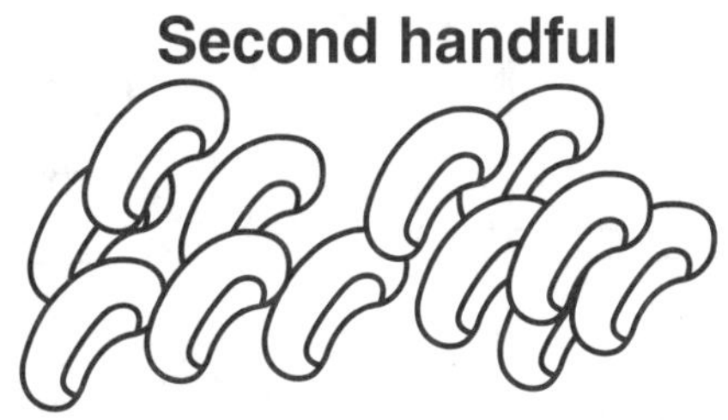

Sort the beans into groups of tens and ones. Place 10 beans in each small cup and then place the cups on the tens side of the mat. Place the remaining beans (if 9 or less) on the ones side of the mat. Record the numbers.

Repeat the above step to get the second set of numbers. Start on the ones side and count the number of single beans. If there are 10 or more, place 10 beans in a cup and move the cup to the tens side. Continue doing this until there aren't enough beans to make a cup of 10. Record how many cups of beans were moved to the tens side in the small square.

Starting on the ones side, count the number of beans and record the number. Count the number of cups on the tens side and record the number.

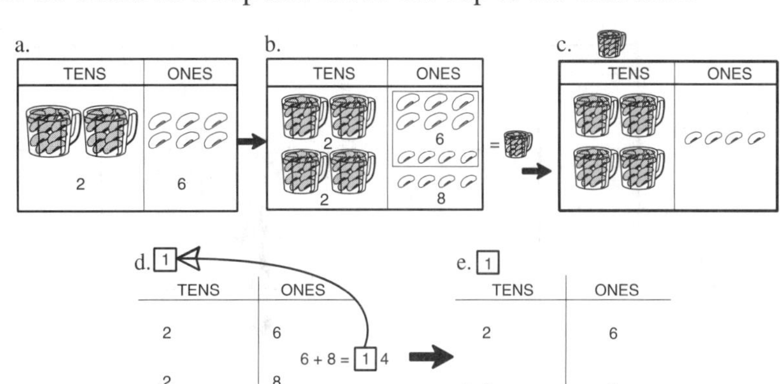

1. tens | ones

$+$

2. tens | ones

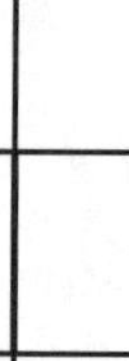

$+$

3. tens | ones

$+$

4. tens | ones

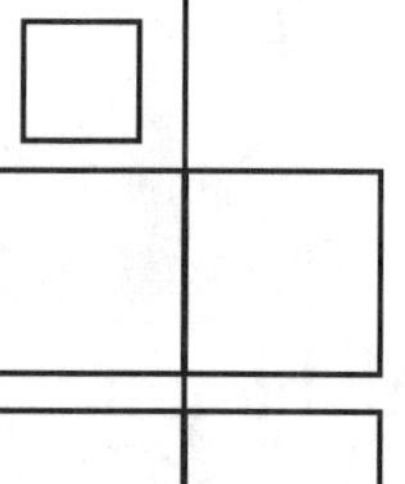

$+$

5. tens | ones

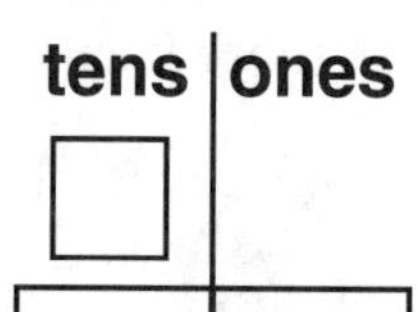

$+$

6. tens | ones

$+$

7. tens | ones

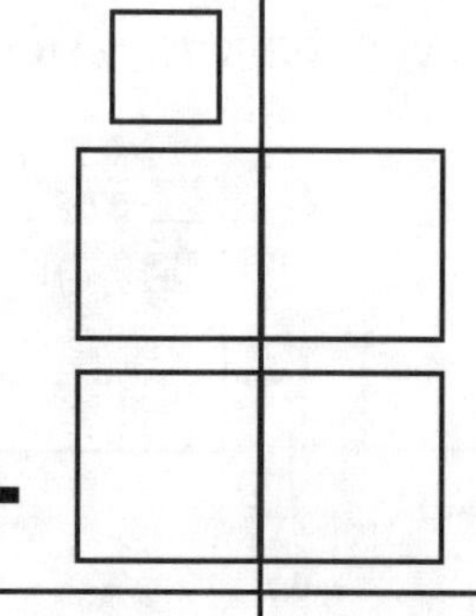

$+$

8. tens | ones

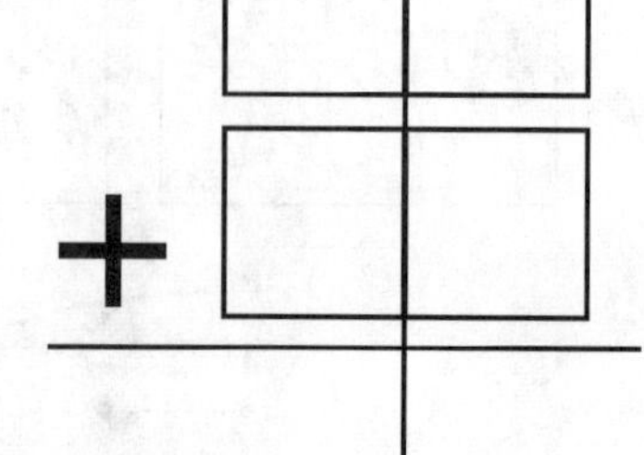

$+$

9. tens | ones

$+$

4 Practice •••• Two-Digit Addition with Regrouping

When adding two digits, always start on the ones side first.

For example,

	tens	ones
	1	
	5	4
+	2	7
	8	1

- The 4 ones and the 7 ones make 11 ones. That is the same as 1 ten and 1 one.
- Remember, there can only be 9 ones on the ones side. So take 10 of the ones and move that set of 10 to the tens side.
- In the small square at the top of the left column (or tens column) record how many tens were moved to the tens side.
- Now add the remaining ones and write the number. Add the tens and write the number.

1.

	tens	ones
	3	9
+	1	3

2.

	tens	ones
	4	2
+	2	8

3.

	tens	ones
	6	1
+	1	9

4.

	tens	ones
	2	5
+	2	5

5.

	tens	ones
	3	6
+	3	6

6.

	tens	ones
	1	8
+	1	7

Solving Addition Problems

Read each word problem. Write the math problem and solve.

1. At the zoo, Alicia fed the monkey 29 peanuts and the elephant 47 peanuts. How many peanuts did the animals eat in all?

   The animals ate _______ peanuts in all.

   $\begin{array}{r} 29 \\ +\ 47 \\ \hline \square\square \end{array}$

2. Robbie watched the zookeeper feed the seals. The first seal ate 12 fish. The second seal ate 27 fish. How many fish did the seals eat in all?

   The seals ate _______ fish in all.

3. The flamingos love to eat shrimp. The flamingos ate 25 shrimp in the morning and 19 shrimp in the afternoon. How many shrimp did the flamingos eat in one day?

   The flamingos ate _______ shrimp in one day.

4. There were 67 kids in the petting zoo. Then 30 more kids came and joined them. How many kids are now in the petting zoo?

   There are _______ kids in the petting zoo.

5. Write your own word problem.

   ____

   ____

   ____

   ____

Challenge

Circle all of the problems with regrouping.

5 How to • • • • • • • • • • • Regroup When Subtracting

Learning Notes

Children continue practicing with the concept of "tens" and "ones." They begin regrouping ("breaking apart" or "borrowing") groups of tens into ones and learn to do subtraction problems with regrouping.

Materials

- sets of tens manipulatives: craft sticks, straws, or coffee stirrers in sets of ten (rubber banded together); multilink cubes or connecting chains (snapped together to make tens); beans or small candies placed in sets of ten in paper cups (Sets of items can be broken apart when regrouping needs to be done.)
- Tens and Ones Mat which can be drawn on a piece of paper (See the example on the right.)

tens	ones

Teaching the Lesson

Introduce the idea of regrouping by modeling how a group of ten can be "regrouped" into ones. For example, show "13" using the idea of tens and ones. Ask the children how many tens and ones there are. (There is 1 ten and 3 ones.) Break the ten into ones. Ask the children how many tens and ones there are now. (There are 0 tens and 13 ones.) Let the children have the opportunity to practice this idea of "regrouping" several times.

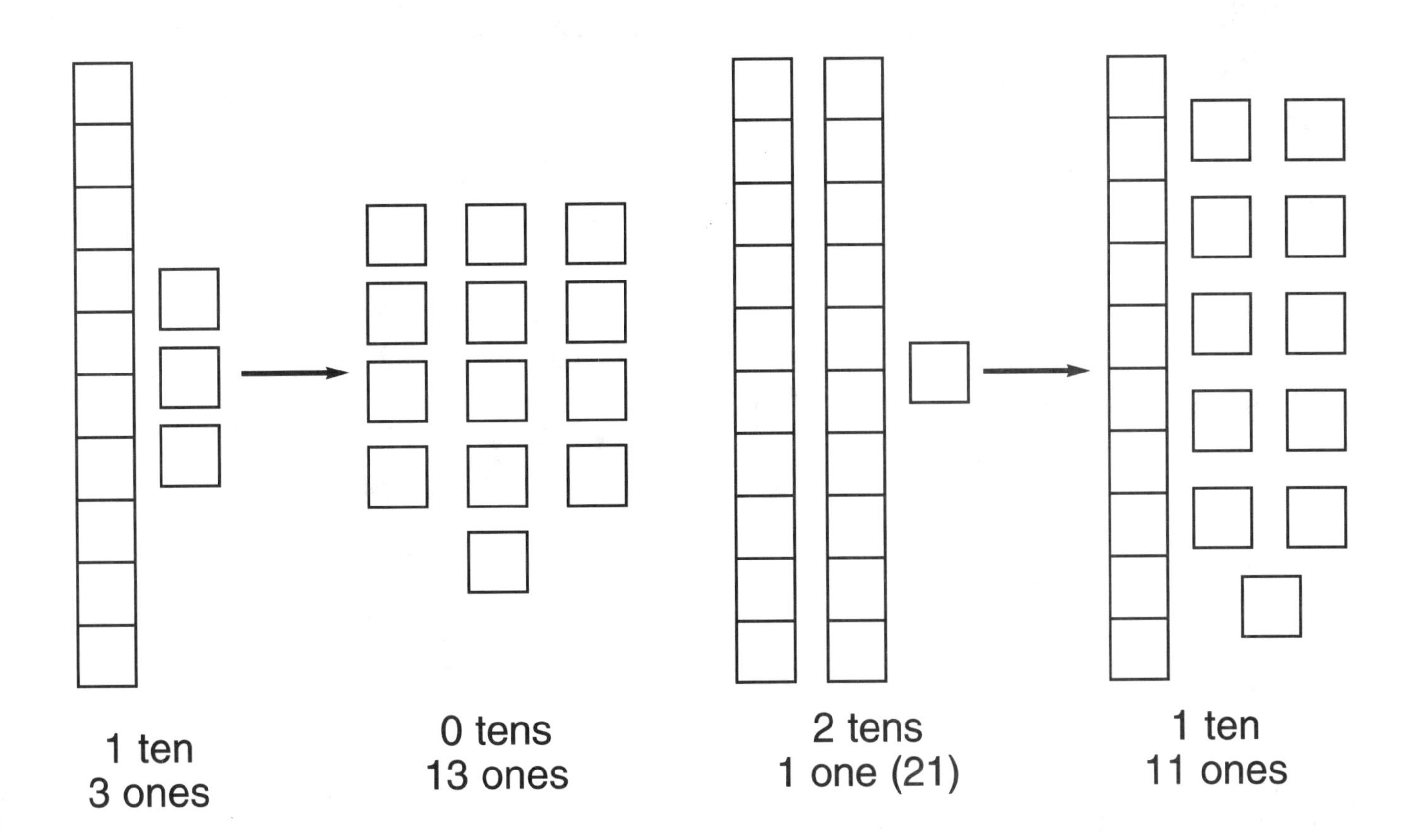

Teaching the Lesson *(cont.)*

Model for the children how to regroup when subtracting one number from another. Remind the children to always start on the ones side.

Remind children that in the math problem, 50 – 29, that 9 ones can not be subtracted from 0 ones. The children will need to regroup ("borrow") a ten from the 5 tens.

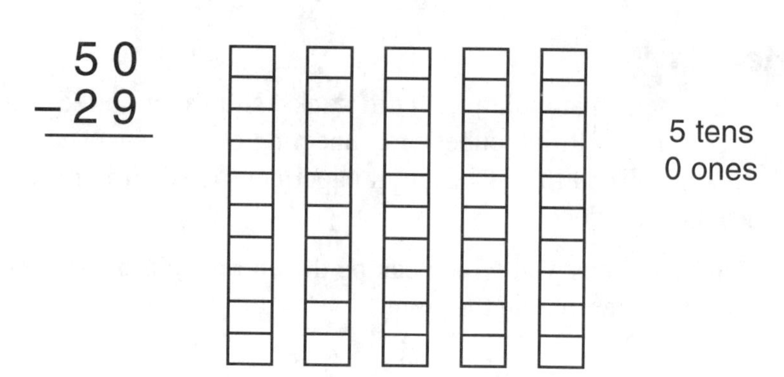

To regroup, the children will subtract a ten from the 5 tens. To do this, change the 5 (tens) to 4 (tens). Move the borrowed ten to the ones side and "break" it apart (show it as 10 ones). Now there are 4 tens and 10 ones.

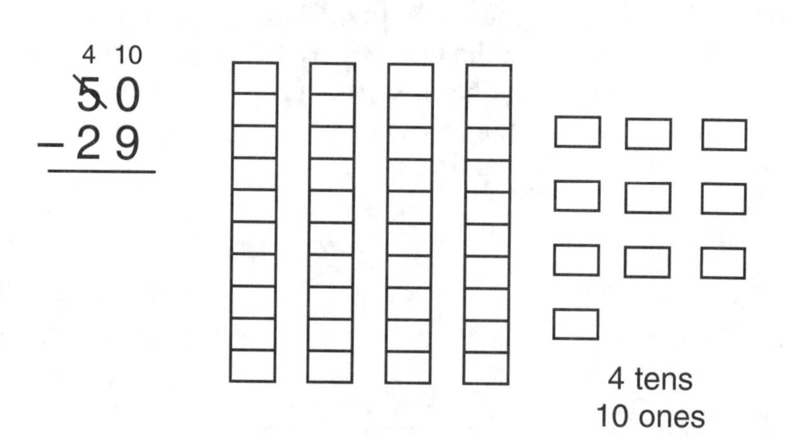

Subtract the 9 ones from the 10 ones and write the answer.

Subtract the 2 tens from the 4 tens and write the answer.

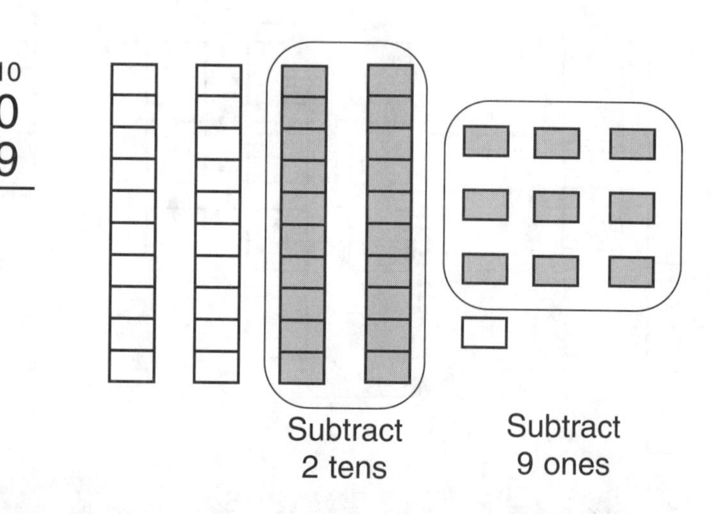

Give the children many opportunities to use manipulatives to do the regrouping before introducing paper-and-pencil work sheets. The more experience the children have in actually going through the process of regrouping, the easier it will be for the children when they tackle paper-and-pencil activities. This will also give the children a greater understanding of regrouping and when and why it is done.

Two-Digit Subtraction with Regrouping

When subtracting two digits, always start on the ones side first.

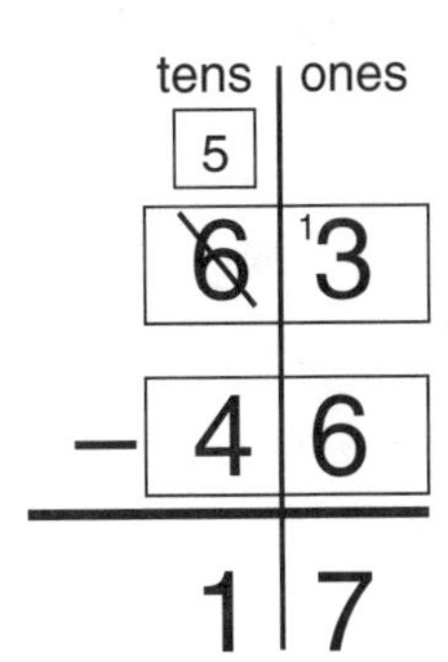

- In the problem 63–46, the 6 ones can not be subtracted from the 3 ones. You will need to regroup.
- Take a ten from the 6 tens. Change the 6 tens to a 5 and move the borrowed ten to the ones side. Add 10 to the 3 ones. Now there are 5 tens and 13 ones.
- Subtract the 6 ones from the 13 ones. Write the answer.
- Subtract the 4 tens from the 5 tens. Write the answer.

Solve the problems below.

1. 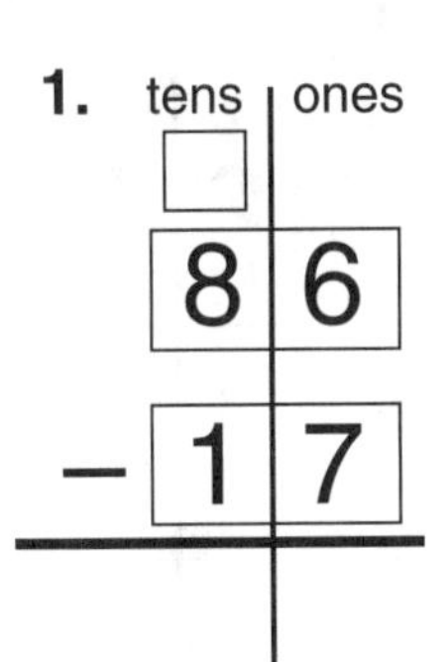

2. tens | ones
   63
   − 45

3. tens | ones
   90
   − 55

4. tens | ones
   71
   − 52

5. tens | ones
   53
   − 28

6. tens | ones
   40
   − 29

Check your work! Add the answer to the minuend (the number being subtracted). The total should be the same as the largest number.

5 Practice • • • • • • • • • • • More Two-Digit Subtraction

Complete the subtraction problems. On some of the subtraction problems, you will need to regroup. Use manipulatives to help you work through each problem.

1. tens | ones

tens	ones
8	
~~9~~	[1]1
− 8	5
	6

2.

tens	ones
7	5
− 5	7

3.

tens	ones
6	6
− 5	8

4.

tens	ones
5	3
− 3	3

5.

tens	ones
3	5
− 1	8

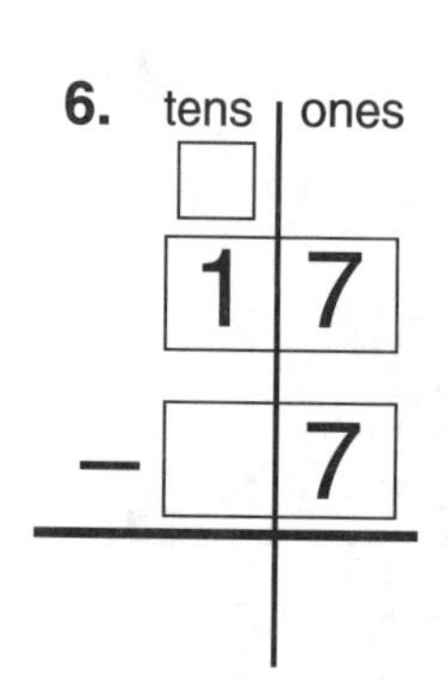

6.

tens	ones
1	7
−	7

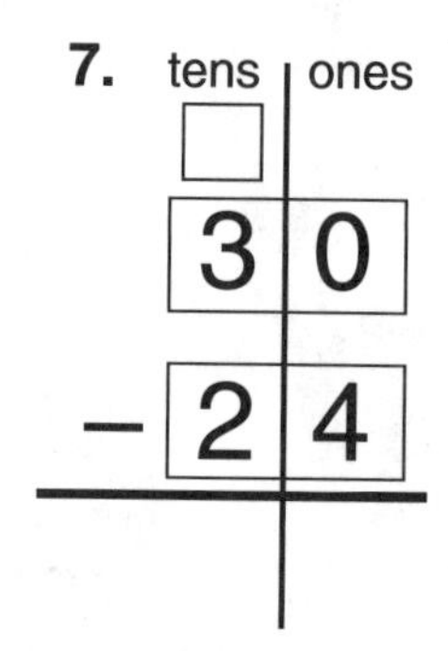

7.

tens	ones
3	0
− 2	4

8.

tens	ones
2	1
−	9

9.

tens	ones
7	9
− 3	9

10.

tens	ones
4	8
− 2	9

11.

tens	ones
8	3
− 7	1

12.

tens	ones
6	7
− 3	8

Challenge

Circle all of the problems where regrouping was needed.

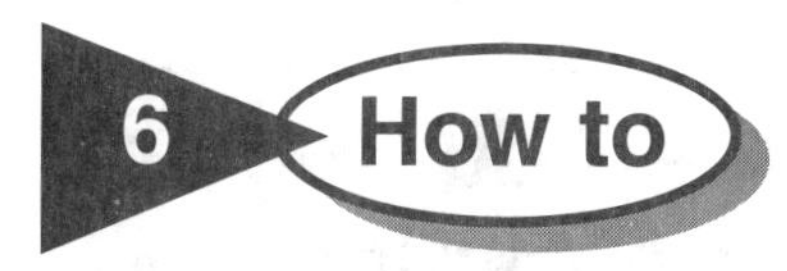

•••• Reinforce Addition and Subtraction

Learning Notes

Children practice adding and subtracting with two-digit numbers. They also practice regrouping and use a calculator to check their answers.

Materials

- number chart to 100 (page 27)
- scratch paper or math journal
- different colors of counters
- crayons
- pennies
- a spinner with six sections: three sections labeled "more than" and three sections labeled "less than." An easy way to make a spinner is to draw a circle on a piece of index paper. Use a paper clip as the "arrow." Place one end of the paper clip in the middle of the spinner. Place the end of a pen or pencil through the paper clip on the center point of the spinner. Hold the pen or pencil in place with one hand and gently spin the paperclip with the other hand.
- beans or other small items
- calculators

1	2	3	4	5	6	7	8	9	10
11	12	13	14	15	16	17	18	19	20
21	22	23	24	25	26	27	28	29	30
31	32	33	34	35	36	37	38	39	40
41	42	43	44	45	46	47	48	49	50
51	52	53	54	55	56	57	58	59	60
61	62	63	64	65	66	67	68	69	70
71	72	73	74	75	76	77	78	79	80
81	82	83	84	85	86	87	88	89	90
91	92	93	94	95	96	97	98	99	100

Teaching the Lesson

Using a Hundreds Chart (page 27): Use the chart for the following activities.

Make 150

The children need to pick three numbers on the hundreds chart that added together will make exactly 150. Each child needs to write down and add the three numbers to see if the numbers make 150. The children can use a calculator to check their addition. If the numbers make 150, the child uses a crayon to color in the three numbers. Then the next player takes a turn.

Variation: Pick another target number and play the game again.

Who Are My Neighbors?

Give each child a penny or a counter. The child throws the counter onto the hundreds chart and adds together the neighbors: the numbers above, below, and to the right and left of the counter. The child writes the math problem down on a piece of paper or in a math journal. Use a calculator to check the addition.

Variation (for 2 or more players): Each child follows the same rules outlined above, but after all players have taken a turn, the spinner is spun. If the spinner lands on "more than," the player with the largest number wins the game. If the spinner lands on "less than," the player with the smallest number wins.

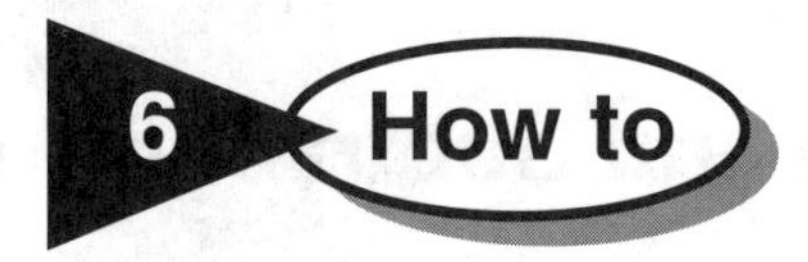

Reinforce Addition and Subtraction *(cont.)*

Teaching the Lesson *(cont.)*

Spill the Beans

Take three beans and toss them onto the hundreds chart.

On a piece of scratch paper or a math journal, add the first two numbers together and subtract the third number. Record the answer.

Variation (for 2 or more players): After each player has taken a turn, spin the spinner. If the spinner lands on "more than," the player with the largest number wins. If the spinner lands on "less than," the player with the smallest number wins.

Addition and Subtraction in a Game (page 28): Use index cards and the game to reinforce addition and subtraction skills.

Materials

- 3" x 5" (8 cm x 13 cm) index cards numbered 1 to 100
- recording sheet (page 28)

Addition and Subtraction Game

Shuffle the index cards and place them in a stack facedown. Take the top three cards. Add the first two cards together and record the answer. Then subtract the third card's number and record the answer. (**Note:** Tell children that if the number on the third card is larger than the sum of the other two, they need to choose another card that is equal to or less than that sum. This will avoid the problem of negative numbers.)

Variation (for 2 or more players): After each player has taken a turn, spin the spinner. If the spinner lands on "more than," the player with the largest number wins. If the spinner lands on "less than," the player with the smallest number wins.

Add and Spin

Follow the directions for making the spinner but label two sections with "ones," two sections with "tens," and two sections with "hundreds."

Take the top two cards and add them together. Spin the spinner. Circle the number in the answer that corresponds with the label on the spinner. Repeat the above steps.

Variation: Take the top two cards and subtract the smaller number from the greater number. Spin the spinner. Circle the number in the answer that corresponds with the label on the spinner. Repeat the above steps.

6 Practice • • • • • • • • • • • • • **Using a Hundreds Chart**

Use this chart to help you add and subtract numbers from 1–100.

- To add 25 + 4, for example, place a marker (or your finger) on 25 and move ahead 4 numbers to 29. (25 + 4 = 29)
- To subtract 48 – 9, for example, place a marker (or your finger) on 48 and move back 9 numbers to 39. (48 – 9 = 39)

1	2	3	4	5	6	7	8	9	10
11	12	13	14	15	16	17	18	19	20
21	22	23	24	25	26	27	28	29	30
31	32	33	34	35	36	37	38	39	40
41	42	43	44	45	46	47	48	49	50
51	52	53	54	55	56	57	58	59	60
61	62	63	64	65	66	67	68	69	70
71	72	73	74	75	76	77	78	79	80
81	82	83	84	85	86	87	88	89	90
91	92	93	94	95	96	97	98	99	100

6 Practice •••• Addition and Subtraction in a Game

Number one hundred 3" x 5" (8 cm x 13 cm) index cards from 1–100. Shuffle the cards and place them in a stack facedown. Take the top three cards. Add the first and second card together and subtract the third card. Write down all three numbers as well as the answer.

1.	2.	3.
36 + 12 ——— 48 – 25 ——— 23	☐ + ☐ ——— ☐ – ☐ ——— ☐	☐ + ☐ ——— ☐ – ☐ ——— ☐
4.	**5.**	**6.**
☐ + ☐ ——— ☐ – ☐ ——— ☐	☐ + ☐ ——— ☐ – ☐ ——— ☐	☐ + ☐ ——— ☐ – ☐ ——— ☐

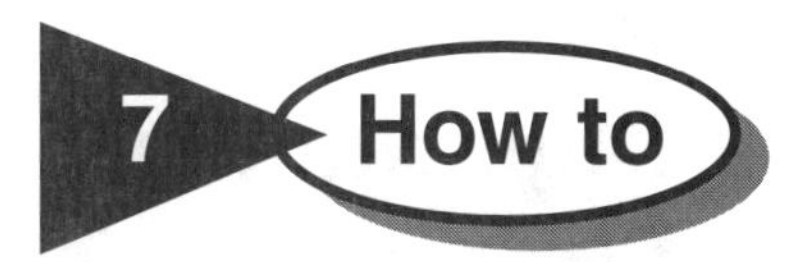

Regroup with Three-Digit Numbers

Learning Notes

The children practice place value to three places: hundreds, tens, and ones. They also learn to regroup to two places with three-digit numbers.

Materials

- sets of hundreds manipulatives: craft sticks, straws, or coffee stirrers (rubber banded together in sets of ten and then rubber banded in sets of hundreds); multilink cubes or connecting chains (snapped together to make tens and then put in sets of hundreds); beans or small candies placed in sets of ten in paper cups with ten cups placed on a paper plate to make 100
- calculators
- 3" x 5" (8 cm x 13 cm) index cards numbered 0–9
- Place Value Mat, which can be drawn on a piece of paper, as shown in the example on the right

hundreds	tens	ones

Teaching the Lesson

Provide the children with many opportunities to practice place value to three places. Using counters and the Place Value Mat, model for the children how to make different three-digit numbers. Give the children many three-digit numbers to practice using the manipulatives.

Model for the children how to regroup to two places. For example when adding,

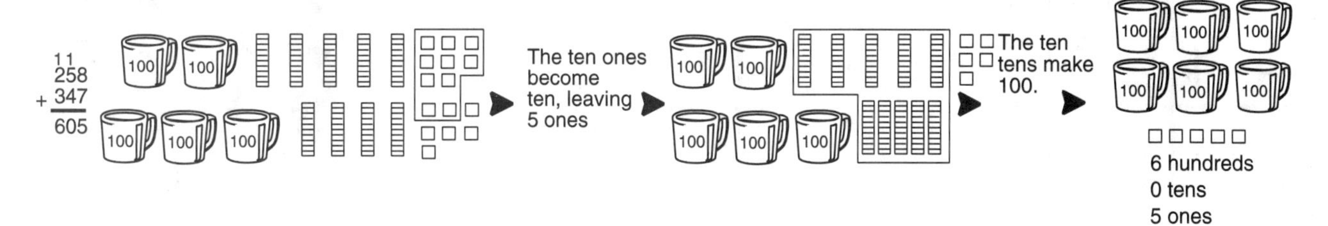

For example when subtracting,

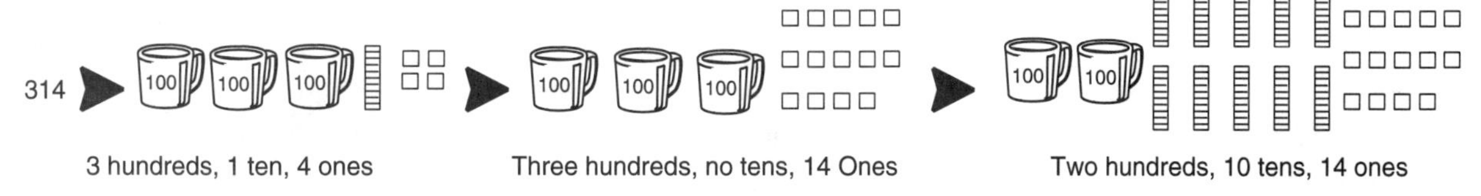

Have the children practice making one number, then regrouping to make another number. Model how to do the practice pages.

Extension Ideas

Use index cards numbered from 0–9 and complete "Writing Math Problems" on page 32.

Shuffle the cards and place in a stack facedown. Take the top three cards and turn them over in order, left to right. Write the three numbers down on the recording page. Take the next three cards and turn them over in order. Write the three numbers down. Then look at the numbers and decide whether to add or subtract. Repeat the steps above to make the rest of the math problems.

Students decide to add or subtract and write the appropriate sign in the circle.

Adding Three-Digit Numbers

Write the answer to each addition problem. Use a Place Value Mat and counters to help you. The first problem has been done for you.

1.

hundreds	tens	ones
1	1	
2	8	6
+ 1	2	9
4	1	5

2.

hundreds	tens	ones
2	9	7
+ 3	0	2

3.

hundreds	tens	ones
4	2	5
+ 1	3	8

4.

hundreds	tens	ones
5	0	7
+ 1	2	6

5.

hundreds	tens	ones
8	0	0
+ 1	0	3

6.

hundreds	tens	ones
7	2	7
+ 1	6	3

7.

hundreds	tens	ones
3	9	1
+ 2	8	7

8.

hundreds	tens	ones
2	1	5
+ 6	0	9

9.

hundreds	tens	ones
3	1	1
+ 1	8	9

Challenge

Circle all of the problems that needed to be regrouped.

Subtracting Three-Digit Numbers

Write the answer to each subtraction problem. Use a Place Value Mat and counters to help you. The first problem has been done for you.

1.

	hundreds	tens	ones
		3	
	3	~~4~~	[1]8
−	1	2	9
	2	1	9

2.

	hundreds	tens	ones
	5	8	5
−	1	6	5

3.

	hundreds	tens	ones
	8	2	5
−	1	8	0

4.

	hundreds	tens	ones
	4	9	9
−	2	0	2

5.

	hundreds	tens	ones
	6	8	8
−	1	7	5

6.

	hundreds	tens	ones
	9	0	0
−	8	7	1

7.

	hundreds	tens	ones
	2	2	5
−	1	6	9

8.

	hundreds	tens	ones
	3	7	5
−	1	5	0

9.

	hundreds	tens	ones
	4	4	4
−	2	6	7

Challenge

Circle all of the problems that needed to be regrouped.

Writing Math Problems

Use the 3" x 5" (8 cm x 13 cm) index cards numbered from 0–9 to make three-digit addition and subtraction problems. Use a calculator to check your work. The first problem has been done for you.

	hundreds	tens	ones
	1	1	
	6	7	9
+	2	4	8
	9	2	7

hundreds tens ones

hundreds tens ones

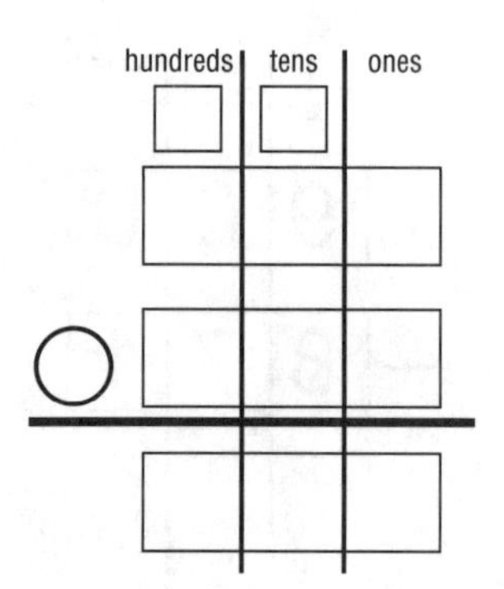

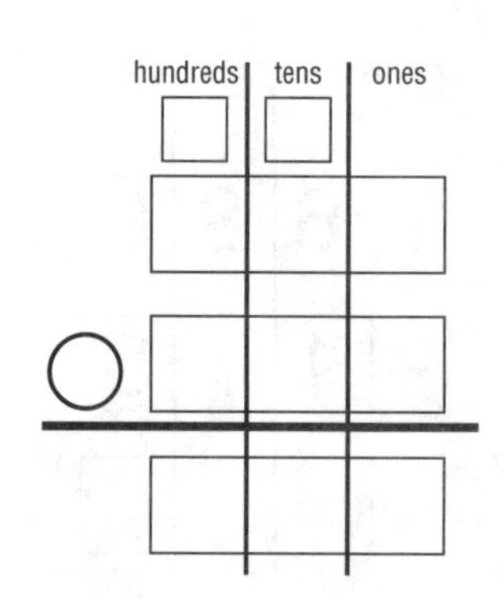

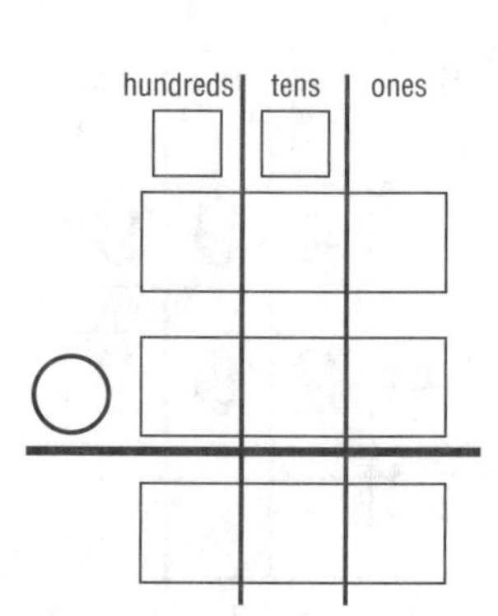

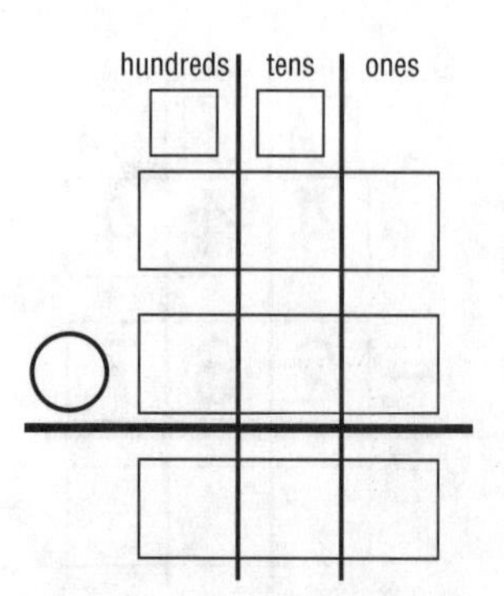

hundreds tens ones

hundreds tens ones

Do you like adding and subtracting numbers? Why? ______________________________

____

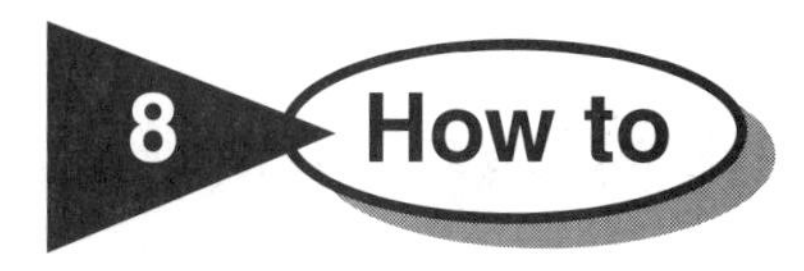

Use Estimation in Addition and Subtraction

Learning Notes

In this unit children practice making estimates and checking their estimates for accuracy. The children also learn to use a ruler. They apply knowledge of tens and hundreds when making their estimates and use a calculator for checking answers.

Materials

- calculators
- paper or math journals
- beans or other manipulatives (counters, multilink cubes, bottle caps, etc.)
- containers of different sizes and shapes
- small cups
- scissors
- rulers
- string
- glue

Teaching the Lesson

Estimating Lengths (page 34)

Before starting the lesson, go over the markings on a ruler with the children. Model how to measure an object by placing the beginning end of the ruler under the beginning end of the object being measured.

Have the children practice measuring items in their desks such as books, pencils, crayons, etc. Check for understanding of how to use the ruler before beginning the practice page.

Estimating Amounts (page 35)

Before starting the activity, model for the children how to fill a cup full of beans (or the manipulative being used). Decide as a group what constitutes "full": all the way to the top of the cup, overflowing the cup, or just below the top of the cup. Model how to count the beans in the cup by grouping the beans into sets of 10 (or 20, 50, etc.) to make counting and recording easier.

Have the children make their estimates before doing the actual measuring and counting of each item. After sorting the manipulatives into groups, the children may use calculators to help in counting all of the manipulatives.

Estimating for Reasonable Answers (page 36)

Before starting the activity, verbally practice making estimates with the children. Present the following example: "I have 50 flowers and I pick 35 more. Do I have about 60 flowers or 90 flowers?" (*90 flowers*).

After completing the practice page, the students may use calculators to check their answers for accuracy and reasonableness.

8 Practice • • • • • • • • • • • • • • • • • • Estimating Lengths

Estimate the length of each item. Record your estimate. Measure each item and record the actual length of each item.

Items to be Measured	Estimate	Actual
1.	______ in.	______ in.
2.	______ in.	______ in.
3.	______ in.	______ in.
4.	______ in.	______ in.
5.	______ in.	______ in.
6.	______ in.	______ in.
7. Cut a piece of string the length of your thumb and glue the string in this box.	______ in.	______ in.
8. Cut a piece of string the length of a large paper clip and glue the string in this box.	______ in.	______ in.
9. Cut a 7" (18 cm) piece of string the length of a new crayon and glue the string in this box.	______ in.	______ in.

8 Practice •••••••••••••••• Estimating Amounts

Look at the cup. How many beans will it take to fill the cup? Write down your estimate. (An estimate is a guess.) Then fill the cup with beans. How many beans did it take? Write down the actual number.

Actual: ____________

Estimate: ____________

Pick out a container to measure. Draw a picture of the container. Write down your estimate for cups of beans and number of beans and to fill the container. Then count the actual number for cups of beans and number of beans and record your answers. Repeat the same steps with several other containers of different sizes and shapes. Write down what you noticed in your math journal or on a separate piece of paper.

Container	Cups of Beans	Number of Beans
1.	Actual: ____________ Estimate: ____________	Actual: ____________ Estimate: ____________
2.	Actual: ____________ Estimate: ____________	Actual: ____________ Estimate: ____________
3.	Actual: ____________ Estimate: ____________	Actual: ____________ Estimate: ____________

•••• Estimating for Reasonable Answers

Fill the Jar

The chart shows how many of each item will fit into the jar. Circle the best estimate for each problem.

Items Placed in Jar	Number That Will Fit in Jar
marbles	268
pennies	223
jellybeans	346
macaroni	182
rocks	137
pencils	78
seeds	515
candies	429

1. pencil + jelly beans
   430   500
2. rocks + marbles
   310   410
3. seeds + candies
   850   950
4. macaroni + pencils
   150   250
5. pennies + macaroni
   380   410
6. candies + rocks
   450   550

Look at the numbers below. Find 2 items from the chart that when added together make about the same number.

7. 200 ____________________ + ____________________
8. 350 ____________________ + ____________________
9. 600 ____________________ + ____________________
10. 700 ____________________ + ____________________

Now Try This: What would be the smallest possible number that 2 items could make? What would be the largest possible number that 2 items could make? Write your answers at the bottom of the page.

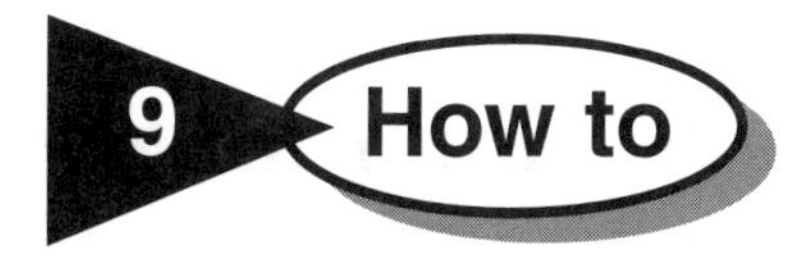

Use Graphs and Charts with Addition and Subtraction

Learning Notes

Children complete charts ("T" tables) to obtain needed information for the activity. They are introduced to the idea of "pattern" and "function." Children manipulate numbers to reach a certain criteria and read a graph to obtain information. They also show through pictures, words, and numbers how they solved each problem.

Materials

- scratch paper
- calculator (optional)

Teaching the Lesson

The practice pages need to be completed in order as information is needed from one work sheet in order to complete the next work sheet.

Before doing the practice pages, go over the concept of "pattern" and "function" with the children (the idea of changing numbers by a specific number). For example, I put 9 peanuts in the box and took 11 peanuts out. I put 5 peanuts in the box and took 7 peanuts out. What is the rule? (The rule is + 2.) Show the children how to use a "T" table to chart their answers.

In	Out
3	5
15	17
1	3

The rule is + 2.

Two is added to each number.

In	Out
7	2
9	4
11	6

The rule is – 5.

Five is subtracted by each number.

Discuss with the children the idea of "estimating" the number of miles traveled (page 38) so that the Wilson family won't travel more than 110 miles in each day. Have the children share strategies they might want to use in order to do this. When possible, have them show how they solved each problem, either through words, numbers, pictures, and/or labels on their practice pages or on a separate piece of paper.

9 Practice •••••••••••••• Using Chart Information

The Wilson family is planning their 4-day vacation. The chart below shows how many miles away each destination is from their hotel. Figure out how many miles it takes to reach each destination and return to the hotel room (round trip).

Destination	Miles	Round Trip
Park	12 miles	24 miles
Zoo	17 miles	____ miles
Waterslide	21 miles	____ miles
Miniature Golf	24 miles	____ miles
Batting Cage	30 miles	____ miles
Picnic Grounds	31 miles	____ miles
Playground	36 miles	____ miles

Now that the Wilson family knows how many miles (round trip) it will take to visit each place, help the family plan their vacation. The Wilsons can travel 110 miles each day. They want to visit all of the places. Make an itinerary (plan) that shows where the Wilsons can go each day and not travel more than 110 miles. Make sure the Wilsons get to visit each place.

Day 1

Destination	Miles
____	____
____	____
____	____

Total Miles: ____

Day 2

Destination	Miles
____	____
____	____
____	____

Total Miles: ____

Day 3

Destination	Miles
____	____
____	____
____	____

Total Miles: ____

Day 4

Destination	Miles
____	____
____	____
____	____

Total Miles: ____

9 Practice • • • • • • Applying Data to Solve Problems

The Wilson family has their itinerary (travel plan) made and they know how many miles they will travel each day. Now they need to figure out how many gallons of gas they will need each day. Their station wagon can travel 18 miles on a gallon of gas. The Wilsons can only buy full gallons, not half gallons, of gas. Figure out how many gallons of gas the Wilson family will need each day. (You will need to use the itinerary you made on page 38 to complete this page.)

Day 1: Total Miles Traveled ________________

Gallons of gas needed ____

Day 2: Total Miles Traveled ________________

Gallons of gas needed ____

Day 3: Total Miles Traveled ________________

Gallons of gas needed ____

Day 4: Total Miles Traveled ________________

Gallons of gas needed ____

9 Practice • • • • • • • Solving Problems Using Charts

At the park, the Wilsons had to wait in line for everything. Look at the chart below. What is the rule? Complete the chart making sure that you follow the rule.

Line	People in Line	Wait Time
1. hot dog vendor	19	_____ minutes
2. seesaw	7	14 minutes
3. slide	11	_____ minutes
4. swing	5	_____ minutes
5. sand box	6	12 minutes
6. monkey bars	9	_____ minutes
7. merry-go-round	15	_____ minutes
8. bounce house	10	20 minutes
9. petting zoo	3	_____ minutes

10. What was the rule or pattern about the number of minutes the Wilsons had to wait in line followed? ____

____

11. How many people in all waited in line before the Wilsons? (Show your work.)

____

12. For what activity did the Wilsons have to wait the longest amount of time?

____

13. What activity had the shortest wait time?

____

Spinners

How to Make a Spinner

- Take three 3" x 3" (8 cm x 8 cm) pieces of cardboard or tagboard. Draw a circle on each piece of cardboard. Divide each circle into 4 equal parts. On the first circle, color 2 sections green and 2 sections blue. On the second circle, color 3 sections green and 1 section blue. On the third circle, color 1 section green and 3 sections blue.
- Make a hole in the middle of the cardboard. Make an arrow out of a piece of cardboard. Make a hole at the end of the arrow.
- Use a paper fastener to attach the arrow to the cardboard.

Spin each spinner 10 times. After each spin, color in a box to show the color to which the arrow pointed.

Spinner 1 (G B / B G)	Green										
	Blue										
Spinner 2 (G G / G B)	Green										
	Blue										
Spinner 3 (G B / B B)	Green										
	Blue										

Spinner 1: The arrow landed on _______________ the most often.

Spinner 2: The arrow landed on _______________ the most often.

Spinner 3: The arrow landed on _______________ the most often.

1. If you wanted to land on green the most often, which spinner should you use? Why?

____

2. If you wanted to land on blue the most often, which spinner should you use? Why?

____

3. If you wanted to land on green half of the time and blue the other half of the time, which spinner should you use? Why?

____

Challenge

Draw a spinner. Use the colors red and orange. Design a spinner that would land most often on red. Tell about the spinner you designed.

10 Brain Teasers

Can You Make 105?

Cut out the numbers at the bottom of the page. Arrange the numbers in the squares so that adding across in each row makes 105 and adding down in each column makes 105. Use a calculator to check your work when you are done.

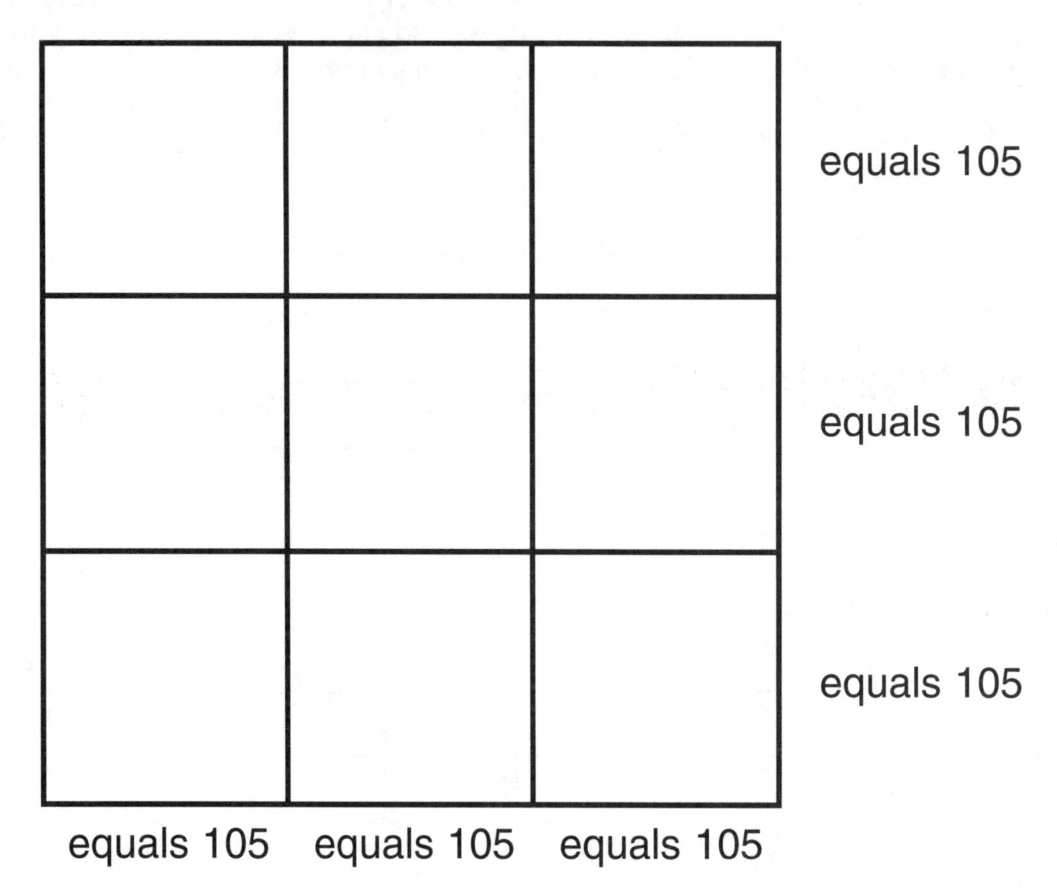

31	32	33	34

35	36	37	38	39

11 Problem Solving • • • • • • • • • • • • • How Many in a Handful?

For this activity you will need a handful of beans or other manipulative. Count how many beans are in your hand. Color in a square to show how many beans are in your hand. Do this 10 times. Record your answer after each handful.

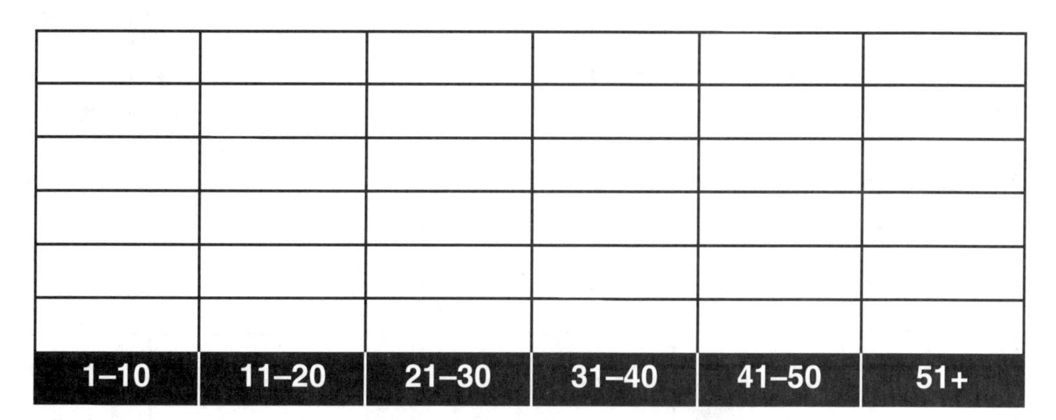

1. Did you pick up the same number of beans each time? ____________________

2. What would be about the average number of beans in a handful? ______________

3. If you had a bigger hand, how would that change the number of beans in a handful?

____

Have a classmate or a family member take 10 handfuls of beans. Record the number of beans in each handful.

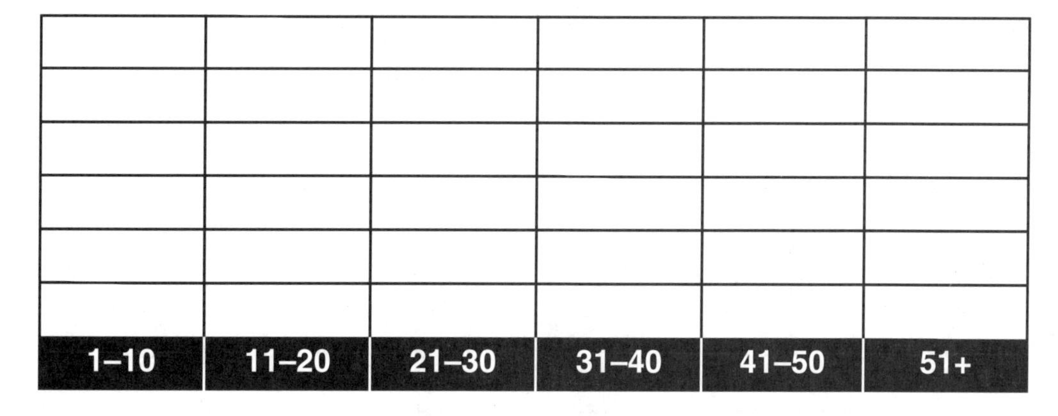

4. Did the person pick up about the same number of beans in each handful as you did?

____

5. What would be about the average number of beans in a handful? ______________

6. How did the other person's handful compare to yours?____________________

7. Write two sentences about the results. ____________________

____

11 Problem Solving • • • • • • • • • • • • • • Pattern Block Fractions

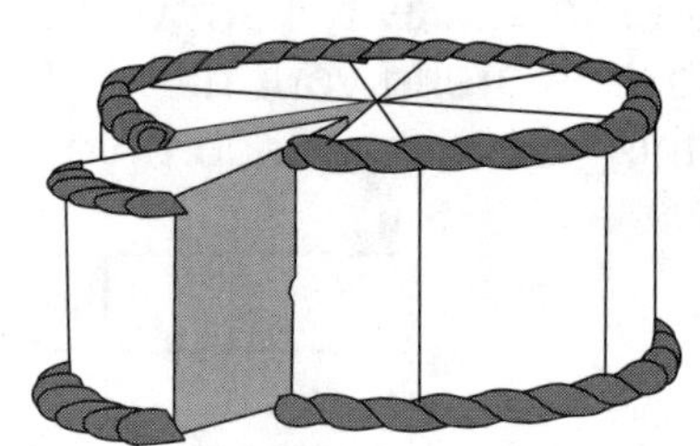

A fraction is a part of a whole item. For example, cut the cake into eight equal pieces. Each piece of cake is $^1/_8$ of the whole cake. If two pieces of cake are eaten, the fraction is $^2/_8$ (two pieces out of the eight pieces).

The top number of the fraction is the **numerator**. The numerator tells how many pieces were eaten or how many parts are needed.

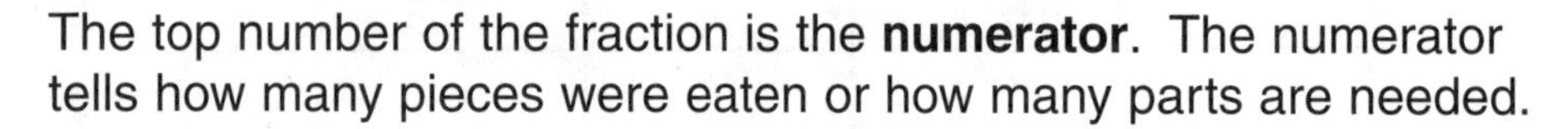

The bottom number of the fraction is the **denominator**. The denominator tells how many parts there are in all. Every fraction has a numerator and a denominator.

Use pattern blocks to complete the chart.

	Hexagon	Trapezoid	Rhombus	Triangle
Triangle	**1.** Pieces to cover shape: ______ Fraction of each shape: ______	**2.** Pieces to cover shape: ______ Fraction of each shape: ______	**3.** Pieces to cover shape: ______ Fraction of each shape: ______	**4.** Pieces to cover shape: ______ Fraction of each shape: ______
Rhombus	**5.** Pieces to cover shape: ______ Fraction of each shape: ______		**6.** Pieces to cover shape: ______ Fraction of each shape: ______	
Trapezoid	**7.** Pieces to cover shape: ______ Fraction of each shape: ______	**8.** Pieces to cover shape: ______ Fraction of each shape: ______		
Hexagon	**9.** Pieces to cover shape: ______ Fraction of each shape: ______			

10. What happened to the size of the piece as the denominator became larger? ____________

Write the fraction to tell how much of each shape is shaded.

11. $\frac{2}{6}$ 2 (parts shaded) / 6 (parts in all)

12.

13.

14.

15.

16.

Drag and Subtract

Learning Notes

Use the information below to guide children as they complete practice page 46.

Materials

- computer with paint software
- printer
- page 46
- multilinking cubes (optional)

Before the Computer

- Copy page 46 for each pair of children involved in the activity.
- Use the computer, an overhead projector, or the chalkboard to model this activity for the children.
- Children should be familiar with the stamp function and the draw and paint tools. Children should also know how to print their drawings.

At the Computer

1. On the monitor, divide a blank document in the paint software.
2. Have children work in pairs (child A and child B) for this activity. Distribute a copy of page 46 to each pair of children.
3. Explain to the children that they will be illustrating number sentences for the number 7 using stamps or the paint and draw tools.
4. Ask child A to stamp 7 copies of a chosen stamp in a row across the bottom of the page.
5. Ask child B to drag any subset of those stamps down to the bottom of the page.
6. Children should then work together using the activity page to create a number sentence representing their actions.
7. Have children drag the stamps back to the top of the page so that all seven are in a row again.
8. On the next turn, child A should drag a different subset of that group down to the bottom of the page.
9. Partners work together to record the appropriate number sentence.
10. Have children continue until all possible combinations of the number 7 have been recorded.
11. Ask children to print their pictures and hand them to you with the activity page.

Extension

- Children can illustrate the combination and write number sentences for other numbers.

Number Sentences

Names: ______________________ and ______________________

Here are the number sentences we found for the number 7.

_______ – _______ = _______
_______ – _______ = _______
_______ – _______ = _______
_______ – _______ = _______
_______ – _______ = _______
_______ – _______ = _______
_______ – _______ = _______

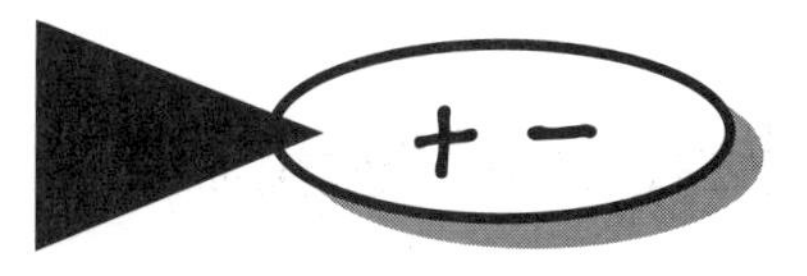

Page 6

1. 2
2. 3
3. 16
4. 17
5. 14
6. 15
7. 10
8. 11
9. 6
10. 7
11. 18
12. 19
13. 8
14. 9
15. 0
16. 1
17. 12
18. 13
19. 13
20. 4
21. 16
22. 10
23. 1
24. 7

Page 7

1. already done
2. 18
3. 15
4. 6
5. 15
6. 15
7. 8
8. 9
9. 16
10. 12
11. 17
12. 13
13. 3
14. 4
15. 0
16. 6

Page 8

1. already done
2. 7 + 8 + 1 = 16
3. 6 + 5 + 9 = 20
4. 3 + 5 + 6 = 14
5. 5
6. 15 – 3 = 12
   12 – 8 = 4
7. 14 – 2 = 12
   12 – 6 = 6
8. 17 – 5 = 12
   12 – 8 = 4

Page 10

1. 14, 16, 18, 20, 22
2. 37, 39, 41, 43, 45
3. 31, 33, 35, 37, 39
4. Counting by 2's
5. 30, 35, 40, 45, 50
6. 86, 91, 96, 101, 106
7. 62, 67, 72, 77, 82
8. Counting by 5's
9. 80, 90, 100, 110, 120
10. 35, 45, 55, 65, 75
11. 73, 83, 93, 103, 113
12. Counting by 10's
13. 200, 250, 300, 350, 400
14. 280, 330, 380, 430, 480
15. 760, 810, 860, 910, 960
16. Counting by 50's
17. 600, 700, 800, 900, 1000
18. 410, 510, 610, 710, 810
19. 575, 675, 775, 875, 975
20. Counting by 100's

Page 11

1. already done
   already done
2. balloon, cloud
   A A B C D A A B
3. snail
   A B B C D D A B
4. 2, 3, 4
   A, AB, ABC, ABCD
5. Patterns will vary.
   Labeling will vary.

Page 12

Patterns and sentences will vary.

Page 14

1. 66
2. 99
3. 60
4. 87
5. 59
6. 58
7. 98
8. 92
9. 49
10. 37
11. 86
12. 88
13. 38
14. 89
15. 91
16. 37, 38, 49, 58, 59, 60, 66, 86, 87, 88, 89, 91, 92, 98, 99
17. 91, 92, 98, or 99
18. 49, 59, 89, or 99
19. 37, 38
20. 60

Page 15

1. 10
2. 1
3. 12
4. 50
5. 30
6. 31
7. 70
8. 33
9. 63
10. 11
11. 48
12. 20
13. 60
14. 21
15. 20
16. 1, 11, 21, 31, 33, 63
17. 10, 12, 20, 30, 48, 50, 60, 70
18. even number
19. odd number

Page 16

1. 15
2. 22
3. 46
4. 22
5. 46
6. 22
7. 60
8. 39
9. 30
10. 60
11. 22
12. 72

Message: I am a math star!

Page 18

Answers will vary.

Page 19

1. 52
2. 70
3. 80
4. 50
5. 72
6. 35

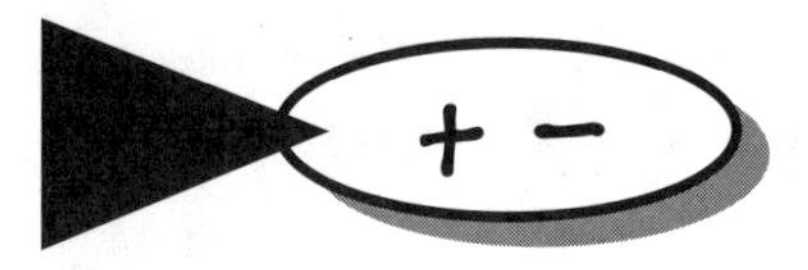

Page 20

1. 29 + 47 = 76
2. 12 + 27 = 39
3. 25 + 19 = 44
4. 67 + 30 = 97
5. Answers will vary.

Challenge: 1, 3, and possibly 5

Page 23

1. 69
2. 18
3. 35
4. 19
5. 25
6. 11

Page 24

1. 6
2. 18
3. 8
4. 20
5. 17
6. 10
7. 6
8. 12
9. 40
10. 19
11. 12
12. 29

Challenge: 1, 2, 3, 5, 7, 8, 10, 12

Page 27

Answers will vary.

Page 28

Answers will vary.

Page 30

1. 415
2. 599
3. 563
4. 633
5. 903
6. 890
7. 678
8. 824
9. 500

Bonus: 1, 3, 4, 6, 7, 8, 9

Page 31

1. 219
2. 420
3. 645
4. 297
5. 513
6. 29
7. 56
8. 225
9. 177

Challenge: 1, 3, 6, 7, 9

Page 32

Answers will vary.

Page 34

Estimates will vary.

Actual lengths:

1. 1½ inches
2. 3½ inches
3. 1 inch
4. 4 inches
5. 2½ inches
6. 2 inches

7–9. Children cut the string to the specified length.

Page 35

Answers will vary.

Page 36

1. 430
2. 410
3. 950
4. 250
5. 410
6. 550
7. pencils and rocks
8. marbles and pencils; pennies and rocks;
9. pencils and seeds; jellybeans and marbles; macaroni and candies
10. pennies and seeds; macaroni and seeds

Now Try This:

smallest—pencils and rocks (215)

largest—seeds and candies (944)

Page 38

Answers will vary.

Page 39

Answers will vary.

Page 40

1. 38 minutes
2. 14 minutes
3. 22 minutes
4. 10 minutes
5. 12 minutes
6. 18 minutes
7. 30 minutes
8. 20 minutes
9. 6 minutes
10. The rule is to double each number to get the answer.
11. 85 people
12. the hot dog vendor (38 minutes)
13. petting zoo (6 minutes)

Page 41

1. *Spinner 2*: The spinner is mostly green and that will increase the chances that the arrow will point to green.
2. *Spinner 3*: The spinner is mostly blue and that will increase the chances that the arrow will point to blue.
3. *Spinner 1*: The spinner has the same number of blue sections as green sections. There is an equal chance of landing on blue as well as green.

Challenge: The spinner should be colored mostly red with a small section of orange. This will increase the likelihood that the arrow will point to red after each spin.

Page 42

One sample solution is noted below.

Row 1: 36, 37, 32

Row 2: 38, 33, 34

Row 3: 31, 35, 39

Page 43

Answers will vary.

Page 44

1. 6 pieces; 1/6
2. 3 pieces; 1/3
3. 2 pieces; 1/2
4. 1 piece; 1/1 or 1
5. 3 pieces; 1/3
6. 1 piece; 1/1 or 1
7. 2 pieces; 1/2
8. 1 piece; 1/1 or 1
9. 1 piece; 1/1 or 1
10. The piece of the item becomes smaller.
11. 2/6
12. 1/3
13. 3/5
14. 1/3
15. 5/6
16. 1/4

Page 46

Answers will vary.